AWARE

AWARENSS

Copyright © 2011 by Gary Louis Warren

All rights reserved ®

No portion of this book may be reproduced, stored in a retrieval system, or transmitted in any form except for brief quotations in print reviews without the prior permission of the author.

2011 Printing

ISBN 978-1-257-82124-2

Published by www.lulu.com

Cover art and Design by Shelby Ann Warren
Edit by Shelby Ann Warren & Dana Armenat Charters
Author: Gary Louis Warren

FOREWORD

This document and the following book are intended as a starting point for an understanding and background to inventions and information that will be of great benefit to a healthy planet and the human race... But there must be a disciplined understanding in their usage. In general, the human ego fights or rejects the concept of mental discipline and order through creative freedom or boredom, but for any healthy biological system a subconscious mental stability and structural discipline is a must. Examples of this lack of subconscious mental discipline or worth are displayed in the pronounced increase in cancer within the human race. As our conscious attitudes, opinions and judgments sift back through to unstabilize the global subconscious.

First off as a disclaimer, I am not a scientist or biologist and don't proclaim to be. Although I am a human being, my terminologies through my experiences are directly related to those experiences and will not fit into many viewpoints. Although my understanding on some points may eventually vary as research dictates, the overall basis that drives the inventions is sound.

Nicolas Tesla, as an inventor, had experiences that Thomas Edison called cheating, but through Tesla's "cheating" experiences, his visions and intuitional knowledge, we have alternating current, radio, florescent lights and more. So what I am expressing below will likely not fit within any current belief or observational structure, just as Tesla's did not. But as we know belief and observational structures are groupings of information, although supportive in many ways, may have limiting and controlling influences that do not take into consideration all possible realities and the totality of the universe.

So with that said, there is a basic misunderstanding in how space and the various atoms and particles in the universe are created, along with continued confusion about alternate realities.

To simply put it there are three main regions in the universe that we must understand. So let me start with the last as the third region, the one we know the most about, as our physical reality; then working backwards through the second then the first.

The third and last region to be created, and most difficult to manipulate, is the region of the universe we call physical reality. This reality permits the practice of observational science; what we can, see, feel, smell and touch. However observational science only allows observation of what is observable, through our senses or instrumentation created from materials within the reference that makes up this physical reality. This is a form of hindsight science that assumes all things were created long ago, and are being manipulated by the unknown, as a gravitational and magnetic force. Although scientists know gravity and the magnetic nature hold galaxies, solar systems, stars, planets, moons, asteroids and all things together, they have no idea where gravity and the magnetic nature comes from, they only know that it is observable.

The second region in the universe and the second region created is nonphysical reality. This region of the universe has basically been denied existence due to its elusive nature to observational science. This nonphysical reality region is made up of energy that is *similar* in content and organization at the subatomic level to physical reality, but is rapidly adjustable in overall groupings as elements, compound and thermal amplitude. The thermal amplitude or radiance of an atom's particle is seen as a difference in color, as some colors are considered cooler in radiance and some are considered warmer in radiance. This nonphysical reality is vaguely detectable from within physical reality through the allusive observation from the human senses, but currently not through physical instrumentation. This may be difficult to understand, but physical reality is directly created from the nonphysical reality region. Our physical reality is created through the three dimensional reflection of nonphysical reality; but not as a reflection, like in a mirror or on a pool of water, like in physical reality. This three dimensional reflection takes place at a subatomic level of the nonphysical reality space and the various atoms and particles that make up the nonphysical reality region. To understand this three dimensional subatomic level reflection one must understand the general makeup of physical reality space and the various atoms and particles that make up the physical reality region.

To start, let's break up an atom into its various parts; protons and neutrons and their internal quarks and their orbiting electrons. This structure in nonphysical reality is quite the same as those in the physical reality region. But the differing parts of an atom in the nonphysical reality region have great differences in thermal amplitude. This thermal amplitude, of let's say a single electron, of an atom will determine the electrons three dimensional refection in the physical reality region. So this same atom in a nonphysical realty region will have more electrons than the amount of electrons in the physical reality counterpart, seeing that some of the electrons in the nonphysical reality do not have the amplitude to reflect. But with the introduction of life to the physical reality region, through the amino acids in DNA, these non-reflected electrons from the nonphysical reality region will increase to the appropriate amplitude or radiance, as their energy passes through the particular amino acid, to reflection in the physical reality. Similar to the change in thermal amplitude or radiance of any given image in a glass mirror that passes through a piece of colored glass before the image reaches the mirror. So atoms from a particular element within the physical reality can be modified into another element from the amino acids alteration of thermal amplitude or radiance to reflection. We see this in new elements that have been introduced to the planet earth through amino acids during the development of physical reality life many millennia ago.

One thing to understand, the reflected atoms from the nonphysical reality are of the same structure but are not of the same type of energy. The particles of the physical reality atom have a much denser structure, decay rate and radiance than those of the nonphysical reality version. The physical reality version of any given reflected atom's particles are of a different colder radiance and compacted version from their nonphysical reality counterpart. The reflected atoms of the physical reality are much different than a reflection we see in a mirror. These reflected atoms take on a life of their own and don't move about with the movement of their non-reflected counterpart. Once they are reflected their relationship is somewhat severed, except for when another particle of a particular atom moves over to reflection from a change in color radiance. It is basically new creation of an atom and space in an altering region of the universe. This whole process gets much more dramatic as there are many levels of reflection and reflections of reflections, so there are many more regions as realities between the nonphysical to the one ultimate physical reality. This complexity also holds true with observational science and instrumentations created from within any given physical region that cannot detect the other reflected regions. All the various regions operate at differing natural laws that are in relationship to their level of reflected density or weight. These alternate realities are somewhat detected through the theoretical versions of science, but all regions as realities originate from the one nonphysical realty.

This nonphysical to physical reality reflection takes place within the first region of the universe. Whereas the nonphysical reality is directly created from this first main region of the universe through its magnetic nature and the physical reality is reflected within the first region of the universe through its reflecting nature. But nonphysical and physical realities do not and cannot become one with this first universal region. They are separate but are more of a separated overlay of the first universal region. This can be envisioned if one was to take four sheets of paper and lay them one on top of the other. Whereas the first and third sheets would be the first universal region and the second and fourth sheet would be the nonphysical and physical reality regions. The second sheet as the nonphysical reality with all possible data added to the sheet would remain separate from the first and third sheet. And the fourth sheet with all possible reflected data added would still remain separate from the first, second and third sheets. But data on the second sheet would reflect through the third sheet to the fourth sheet. Although within our universe there are many more sheets as reflected physical levels of reality, but with this basic understanding I am trying to identify the three main regions of our universe and not the totality in universal complexity.

This first region of the universe is much different then what we think of as a physical or a nonphysical space as a reality. This region of the universe is what we might relate to as a gravitational and magnetic force that has no space as energy and time, but has a reflective nature on the boundary between this region and a nonphysical and physical reality spatial regions. We can more appropriately understand this first universal region if one were to consider this first universal region as what we think of as a black hole. The black hole in space is separate from all possible nonphysical and physical realities, but is the underlayment to all space energy and time. A black hole could be considered as a hole in the fourth sheet as our physical reality region. And the event horizon would be energy made up of all possible nonphysical through physical reality regions as this hole passes through all sheets of reality as the nonphysical and all physical reality regions. But there are black holes that do not reach up through all the varying physical reality sheets as regions but would be observed as an accretion disk on those sheets or regions where the black hole does not penetrate. This first universal region is also glimpsed throughout space and more between galaxies but without the pocketed gravitational as a magnetic force that forms the event horizon energy border. So this first universal region, as with all universal regions, is separate from one another but overlaid on top of one another. And the first universal region has a magnetic and reflective nature that creates and manipulates the nonphysical and all physical reality sheets or regions.

So if we had the capacity to move into a black hole on our way to the first universal region we would be, in essence, traveling through all the varying reality sheets as regions. This same effect of traveling through all the varying reality sheets as regions happens within every active synapse in the brain. The active synapse accesses the first universal region by ripping a hole in all the universal regions or sheets. So the physical through nonphysical brain can access the first universal magnetic and reflective region and vice a versa. This hole created by the activated synapse also allows the nonphysical reality energy, as non-reflected energy, to pass to the physical reality region; that allows this nonphysical reality energy to travel throughout the human nerve system and into the physical reality space. This nonphysical reality energy, that is traveling through the human nervous system and ultimately through the amino acids in the DNA, forces the reflection of ingested compounds (food) to make new compounds that we see as growth or refer to as life.

But as the multiverse brain travels up and down through all the varying reality sheets as regions, from conscious to subconscious in the nonphysical reality, created through the hole formed by the activated synapse, now within all the regional brains we access information from all the differing reality regions as we sleep. Some of this information can be retrieved as dreams, intuitions or visions that may have nothing to do with one's current universal region. Whereas the first universal region would be a clear or see-through region that has a magnetic and reflective nature, this could be considered a clear region or medium that makes up the one universal mental component that is accessed through an activated synapse. So this would make the first universal region a magnetic and reflective region or a Clear Mental Medium.

The book AWARE works forward, starting from the development of the first universal region in detail and up through a detailed description of the second nonphysical region's creation and third, as creation through reflection, as the physical region. This book also suggests understandings for long forgotten inventions that can be recreated for beneficial day to day usage.

With this information is there a possibility of creating a manned vessel that can act as a large synapse and punch through to the first universal region and back again to a new location in an instant? Think of all the benefits of understanding how the human biological system is created and supported from the nonphysical region, through unwavering subconscious mental stability, up through a DNA organized reflection. What if we can create a large synapses like conduit (I have a working prototype) that increases the flow of nonphysical reality energy into the human nerves system and with a disciplined mind can we regrow severed limbs or damaged organs and will it elongate life to 900 + yrs like the ancient text proclaims, who has the answer to these research questions. Could we create a channeled like nonphysical energy flow through a DNA or DNA like medium and create new compounds in an instant? If we were to use the appropriate DNA sources to flow the nonphysical energy through we can create sustenance (food) or building materials without waiting through the long growing and harvesting process. What if a synapse like energy conduit was built and that energy flow was attached to a simple device (I have a prototype) that would radiate that nonphysical energy flow into the surrounding atmosphere and slightly alter those surrounding physical atoms into a denser compound. This would instantly create a kind of energy shelled to protect structures from extreme weather conditions, cars and people from serious collision, maybe even space vessels from decompression, and more... Or what if we built a synapse like energy conduit and were able to convert that nonphysical energy flow into a physical electron flow as direct or alternating current at any point in space (is this what Tesla had envisioned but was missing the fundamental understanding to accomplish)... Should we as a people look into these maters?

PREFACE

What's in this book is instructive background information as inserted into a provisional patent application. This background information is the foundation of why the provisional patent invention works as tested and how this same principle can be extended to a multiple of devices.

This invention background information is a direct guide, to continuous creating of the universe up through to continuous creating of the human race; and not intended for comparisons to current belief based or limiting observation based ideologies. What it is; is what it is and what it is; is a direct interpretation of actual ongoing experiences and events...

The human race as a whole is still regretfully struggling as to who and what we are; and the unmitigated mentally stabilizing responsibility we own in that regard.

It is my intention and hopes to gain enough interest and capital through the sale of this book or find investors, to finance these inventions for the betterment of the human race and the world we live in...

INTRODUCTION

I am not a writer or man of words per say; this book is more of a document that explains in concise detailed descriptions the underpinnings of the universe; that eventually lead to the continuous creating of the human race and a working invention. As well as understandings that can be applied in differing ways.

To explain how and why this invention works one must understand continuous creating of the human brain and how it is continuously accessed by the aware living spirit. This very access point or hole creates a border between the space energy realms, where an unlimited electromagnetically charged border can be drained for consumer use. Much like the synapse of the human brain drains this energy through the human nervure system.

This book may be a tough read for some; but to explain how and why this apparatus works one must understand the basic principles in the continuous creating of the mentally stable and responsible universe; through to the accessible physical reality we accept.

"For clarity; it must be understood there are three forms of space-energy creation:

- First creation, as non-reflected space-energy…
- Second creation, as reflected and re-reflected space-energy, as re-creation…
- Third creation as creativity, as organized manipulation of first and second creation…

It must also be understood the human race as we know it, mainly identifies with Third creation, as in the arts, as design, engineering, science, literature, etc. These forms of creation are more associated with creativity or creative manipulation and study of space-energy, and not actual creation or re-creation."

ORGANIZED PATTERNS of SPACE-ENERGY 1

In essence the known universe consists of *three basic understandings*: the energy filled spaces (galaxies); that are populated with black-holes; which are surrounded and interweaved by a transparent or unknown Clear-medium.

1. Energy is expanding and/or decaying spaces or mediums, which are dictated by varying periods of activity defined as time. *ALL space is energy of some form.*

2. Black-holes are *positional* within spatial-energy and timeless, or *void of all spatial-energy*, and through human observation assumed manipulator of spatial-energy.

3. Clear-medium[1] as an *absorbent or vacuum-like environment*, with a *controlled projective or compacting force*, that has a *reflective nature in its relationship to spatial-energy*, but *singularly void of spatial-energy and black-holes*.

In moving past the one medium that defines space and time (spatial-energy), we are left with black-holes and the clear-medium. Black-holes are positions or locations identified from within spatial-energy; even though timeless, they are directly tied to the development of spatial-energy and do not exist directly in the clear-medium without a spatial-energy barrier. We are now left with the *clear-medium* as an absorbent environment, non-position definable but still definable by *Place*. Seeing this medium is timeless (no spatial-energy), we can still

---

[1] Clear-medium defines a place before what is termed, space-time, dark-energy, dark-matter or black-hole, and can be loosely considered as black as referenced in black-hole.

speak in the terms of before and after. Before the clear-medium, or outside of this absorbent environment, is unknown and can be considered black or nothing. This un-comprehendible UN-known state is still conceivable; by the one and only definition as *Nothing*.

The concept of Nothing *as a place* is basically understandable, but not truly comprehensible from within spatial-energy; but nevertheless it maintains a real conceptual presence before the clear-medium. From this *conceptual presence*, the absorbent nature or clear-medium is driven into its *conceptual place*. The concept of Nothing and its conceptual presence drives its actual *place and position*, in regarding the rule of conceptual events; *concepts only exist through conception*. The absorbent nature is the real *positional conceiving* or absorbing environment, which can be referred to as the clear-medium.

This clear-medium's absorbent nature can be definable as the basic development of *Aware* (not awareness, not self awareness, not sentient awareness - just aware). The mere concept of Aware absorbs spatial-energy. We are aware of the universe, time and space, an ant or a grain of sand and we absorb, focus, reflect and project on those concepts. If contemplation of anything other than the state of Aware takes place, spatial-energy is created, manipulated and absorbed. All information is spatial-energy and Aware absorbs it; but Aware is not energy, it merely creates and manipulates it. Aware is an *absorbent, focusing, reflective, projective force* that creates and manipulates spatial-energy. Reflect on just the absorbent nature of AWARE; this confusingly could be called meditation, but that definition alone is created and manipulated spatial-energy as information, and comes from within the body of spatial-energy.

The nature of Nothing has no density, no dimension, it has no duration, but it does have a real theoretical place; its place is where its place is and its place is the conceptual presence and position of Nothing. From this one place, as the *conceptual-establishment* of the one real *theoretical-position*, develops the beginnings of an absorbent, focusing, reflective, projective nature, or *theoretical-environment*. This theoretical environment is the aftermath or unacceptable and undeniable concept of Nothing's real position. This clear-medium or *absorbent-environment* is the neutralizing force of the conflicting concept of Nothing, to eliminate its *conceptual-position*.

The development of the theoretical absorbent environment or clear-medium, as an aware environment, is forced into *Being*; following the process rule of *conceptual-events*:

1. All concepts have a place.

2. For any place to have its position, its position must be established.

3. No position can be established without conception, or absorption of information.

4. For information to be absorbed, the position of that information must be identified or focused on.

5. Information is not absorbed without movement, as acknowledgement, through reflection of that information.

6. For any position to be established through an absorbing focus, that position is amplified.

7. For any position to be projected on, that position is compacted; even if the conceptual intent is to neutralize that position.

8. Developmental processes only exist through the expanding creation of mental concepts.

So in essence, the theoretical position is truly Nothing and is manipulated by the theoretical environment, or for the lack of a better term, a clear-medium with the ability to create and manipulate its positional concept. The theoretical environments absorbent nature focuses on its one place, projecting its state of Being as Nothing to neutralize its real conceptual position. The conceptual position of Nothing cannot be neutralized through the building of conceptual events, as the *absorbent, focusing, reflection, projection process*.

This process inadvertently intensifies and amplifies the theoretical position of the theoretical environment, and develops the first one-dimensional event horizon. This event horizon is the conceptual point of movement as acknowledgement or separation, as energy filled space between the two concepts (place and position). A depleted barrier as a rip in the fabric of Nothing, or an electromagnetic field, as a depletion of the theoretical environment, or electromagnetic space; as termed dark-energy, but is more clearly termed *non-reflected-energy*; followed by the absorbent, focusing, reflective, projective nature, as the clear-medium or Aware.

As there is no true separation between the absorbent, focusing, reflective, projective nature and place, as the clear-medium and its position, and the interlude, as depleted area, as electromagnetic filled space, as non-reflected-energy; here the state of aware moved to a state of awareness, or value-filled aware as aware in motion, or an aware gestalt of amplified non-reflected electromagnetic spatial-energy. In other words, the clear-medium or Aware is clouded with value-filled information. This conceptual process can be considered a one-dimensional realm; rooted in a state of Being, established by the theoretical position, and its testimonial information as space energy, and a place where the one real position is identified from, as the theoretical environment.

As the *positional focused, absorbent, reflective, projective nature* is moved to an infinite static state, the absorbent nature's conceptual focus is naturally altered, away from the one static position to the opposite, as alternate or conceptually opposing backdrop. Here the focused, absorbent, reflective nature manipulates its positional focusing absorbent characteristic; from an infinite inward positional focused absorption, through a neutralized static state, to an infinite outwardly focused absorbent state; while maintaining its positional state of Being. This outwardly focused absorption elongates and amplifies the electromagnetic field, or event horizon, from its anchor at the positional threshold. In this state of awareness of the one all-encompassing backdrop, the outwardly focused absorbent nature naturally falls to one side, and ultimately circularly encompasses the one position; establishing the first circular event horizon: the first true black-hole. Here we are moving into the concept of a two-dimensional realm, but with no reference to a concept of size.

As the outward focus turns static and the elongated event horizon begins the natural electromagnetic energy decaying process, an inward compacting projection is developed and intensifies. The elongated event horizon of the black-hole begins to fold or collapse, pushed past its natural decay rate, and naturally falls away to one side of the black-hole. This move alters the black-hole's encircling event horizon to a translucent, light gray cone of electromagnetic energy: the first true wormhole. This moves the theoretical environment of Aware through awareness to the conceptual beginnings of a three-dimensional realm, but again, with no reference to a concept of size. As the outward focus continues and the inward projection intensifies, the theoretical environment begins to collapse and the cone shaped wormhole's electromagnetic energy shell folds unto itself; changing in shape to a cylindrical tube, and at its midpoint, increasing in size with a yellowing thermal reaction. This folding affect can be considered as more of a bunching up of electromagnetic energy, pushed past its normal decay rate, creating the thermal reaction.

The shift from the static outward focus, to an inward compacting projection, is brought about through the start of electromagnetic energy's natural decay process. This decaying process is reflected on as a *timed duration*, as energy moves in a backward decaying motion to the one event horizon's threshold. The decaying movement is the alternate of the value-filled amplification process, created through the outward focused absorption. This decaying energy is the very early stage of the one body of Aware's actual conception of its *energy in motion*, triggered by the decaying energy as *non-value-filled energy in motion* or e-motion. This was reflected on as *one motion* or *oneness* (electromagnetic energy in decay), and as Aware experienced it as a prolonged duration of *oneness*, or as a moving sensation of loneliness. This prolonged backward energy decaying motion of loneliness, alters the one body of Aware's inward projections intensity, to eliminate this *ALL consuming*, only *oneness*, conceptual state of Being; while the outward focus was maintained.

This motion of the forced collapse of the decaying electromagnetic energy shell created a value-filled thermal energy reaction, which naturally altered the inward projection of the outwardly focused, absorbent, reflective nature, and re-established and amplifies the static wormhole. The shell of this wormhole is now laced with thermal energy.

Once again in the infinite moment, the focused, absorbent, reflective nature of Aware stabilizes and turns static, and once again the electromagnetic energy event horizon begins the natural decaying process. Once again its compacting projection is turned inwardly, but in this instant, the spatial-energy decaying process is changed by the altering decay rates of the two forms of energy (thermal and electromagnetic). Now the inward compacting projection intensifies in a rhythmic nature, sending a pulsating collapse in the electromagnetic shell of the decaying wormhole tied to the event horizon.

This decaying energy is the very early stage of *non-value-filled energy in an altering state of motion* or e-motions. This was reflected on by the one body of Aware as one overall prolonged motion or *oneness* (electromagnetic energy in decay), laced with cooling thermal energy or *coldness* (thermal energy in decay), and as Aware experienced it as a prolonged duration of *one-coldness*, or as a moving sensation of lonely-despair. These prolonged backward motions of lonely-despair alter the inward projections intensity, to eliminate this *ALL consuming*, only *one-coldness*, conceptual state of Being; while the outward focus was maintained.

Here, multiple rings of thermal energy are created through the decaying wormhole's pulsating collapse and continuously increasing in form, as the inward projection of Aware intensifies to its infinite, neutralizing state. In this process the multiple ringed thermal reactions continue to expand to their individual centers, ultimately forming spheres of thermal energy with an electromagnetic skin or energy shell. These translucent, yellow spheres of thermal energy are the first forms of *matter*; as termed dark-matter, but more clearly termed *non-reflected-matter*, which are created as the inward compacting projection continues to its infinite state. Now the wormhole, laced with thermal spheres, begins to waver or fold from side to side being pushed past its natural decay rates.

Here the wormhole breaks away from the event horizon, and is instantly compacted or weaved into a sphere shape around the black-hole; vaguely comparable to equally spaced knots on a string wrapped around a ball. But the knots are translucent yellow spheres with a diameter matching the translucent, light gray tube or wormhole, and are pushed or bunched into a tight ball shape rather than wrapped.

At this point, the focus of the absorbent reflective nature or clear-medium is naturally altered inward, towards the value-filled movement creating the ball of energy, and the projective nature once again turns to a static state. Here the focus of the clear-medium, or Aware, begins the absorbent and reflection process on the presented information or energy ball. But after a momentary delay the exact pattern of translucent yellow spheres, in the ball of energy, is reflected or mirrored within the absorbing clear-medium; with one major alteration. The reflected pattern of non-reflected-matter is made up of compacted flickering white points, or as *reflected-matter*, with no apparent reflection of the active wormhole. This value-filled movement attracts the focus of the absorbent reflective nature.

The mirrored pattern of thermal energy spheres, or *non-reflected-matter*, would be considered a reflection of non-physical spatial-energy, and forming the reflected or physically mirrored three-dimensional spatial-energy, as *reflected-matter*. Here it seems that only thermal energy is reflected and altered within the physically mirroring absorbent medium. But actually physically mirrored thermal energy is altered into compacted thermal energy points, with an elongated decay rate, and the electromagnetic energy shell is not readily reflected or physically mirrored. Here the clear-medium, or Aware, absorbs non-reflected spatial-energy or information, and that information is acknowledged and reflected within.

The automatic-like change in focus of the absorbent nature with any movement is the natural beginnings of conscious awareness. This adjustment of focus, controlled direction and intensity of absorption, and purposeful selection of information, can be considered the first steps in the development of universal subconscious self-awareness, or choice driven creative curiosity. At this point the state of self-awareness intentionally moves its focus into the physically mirrored environment, to reflect on this newly created, absorbed and reflected information. Here the clear-medium, or Aware, within the infinite moments, absorbs and reflects on all presented patterns of spatial-energy or information.

Once again the theoretical environment moves to a static state; and once again the inquisitive focus of the absorbent reflective nature of

Aware is naturally turned outward, to the unknown all-encircling conceptual backdrop. From here the nature of Aware or clear-medium intensifies its inward compacting projection, past the multiple spatial-energy decay rates; while maintaining an outward focus. This intensified chosen compacting projection of the clear-medium, on the one-position or black-hole, is the gravitational force that drives all energy and matter, in the space filled theoretical environment, back to its position or core black-hole. These are the beginnings of what would be called the interweaving of non-reflected-energy and non-reflected-matter, reflected-energy and reflected-matter, or non-physical and physically mirrored spatial-energy; as the massive collision of the energy matter rich fields, active and physically reflected, are driven into the one black-hole.

As these energy fields are forced together under the massive compacting or gravitational force, developed by the infinite inward projection of the surrounding clear-medium, all spatial-energy is infinitely compacted around the one black-hole; creating an intensified, multiple flavored, mixed thermal reaction. This inward projection of Aware naturally turns static, triggered by the multi energy flavored value-filled thermal reaction, forcing an instant outward expansion within the clear-medium. This is what is known as the Big Bang; an infinite expansion of all forms of energy and matter within the backdrop of the absorbent clear-medium or Aware, and only populated with positional black-holes as the clear-medium focuses its real positional concept; creating newly rooted universal positions as black-holes, bordered with created spatial-energy as an event horizon.

An expanding and/or contracting clear-medium mixed spatial-energy filled environment, or universe, coexisting with the clear-medium and multiple, non-physical through physically mirrored, patterned reflections of spatial-energy. Aware, or the clear-medium, is apparent from within spatial-energy through the effect it has on spatial-energy;

and seen as a controlling expanding state surrounding the universe; or balancing and compacting spiraling galaxies with central, massive black-holes that establish their universal positioning.

In the one-universal outward expansion, supported and amplified buy an inquisitive outward absorbent focus, all possible energy transformations are created within the one massive spatial-energy implosion/explosion. And within this universal shell of spatial-energy, an irregular energy mixture is developed, as eight overlapping, flavored or lopsided energy pockets; with natural patterned energy flows through the spatial-energy envelope, which is interweaved and manipulated by an absorbent, reflective, projective or magnetic nature. This value-filled, constant energy flow, through numerous flavored overlapping segregated spatial-energy pockets, back to the one position; creates a spiraling effect of spatial-energy around the universal center or positional black-hole.

The clear-medium, or Aware, is a timeless absorbent, focusing, reflective, projective nature, with the ability to positionally focus and directionally project that nature in similar or opposing directions or remain neutral. The clear-medium, as the backdrop or scaffolding of all there is; while black-holes are created through the clear-medium's conceptual identification of its real theoretical positions, through focusing and projecting that concept. So in essence, the clear-medium and black-holes are of the same nature; where black-holes, not necessarily circular or completed, are more neutral and positional and bordered by an event horizon as spatial-energy; and the clear-medium as more of a place from where its position is identified and projected from. This compacting projected, or magnetic-push, of the clear-medium is experienced as the gravitational pull or force, created by the black-hole but in essence is the exact opposite; whereas an absorbent or magnetic-nature within a black-hole can and does create the same gravitational effect on the space-energy surrounding it.

Understanding the nature of Aware and awareness is much different than what humans would call psychologically cognizant, thoughtful sentient, self awareness. This type of awareness is massively clouded with differing strict organized patterns of purposefully accessible information; as all forms of moving spatial-energy, all be it harmonious or conflictive, that is so integrated with the clear-medium, that the energy information shell, or ego, is more seen as the Aware itself. This is where the massive spatial-energy confusion as awareness, and realistic energy patterned illusion begins. These pages of organized text create a similar illusionary shell of information.

The clear-medium or Aware: a manipulative variable absorbent, focusing, reflective, projective medium; the fundamental creator, foundation, and coincident backdrop of all spatial-energy.

The positionally defined clear-medium or Aware backdrop; can selectively operate at differing projective rhythms and characteristics; creating, manipulating and reflecting spatial-energy into organized patterns, that begin in non-reflected spatial-energy, and are ultimately directly influenced from within that energy's natural rhythmic decay rates. These rhythmic projections and patterned spatial-energy formations from the positionally defined aware medium, to the more physically reflected energy planes, require a greater or more aggressive methodical approach; and in some instances are aided through energy movement and refraction, into the more physically mirrored reflections, through crystalline materials as chemical compounds; and in the most specialized form as DNA. And to that regard; the actual positionally defined envelope of the clear-medium, or state of Being, in some instances is projected into more physically oriented reflections, through firing synapses. The DNA refracted energy, within the synapse, develops an energy vibration, as a reverberation between the nerve cells or within the synapse. This energy reverberation develops a spatial-energy depleting characteristic, in physically reflected environments, and punches through the interlaced energy rich fields. Creating an active energy depleted string or *clear-string*, within the synapse to the coincident clear-medium or aware backdrop; allowing a projected energy flow from non-physically reflected environments. The positionally defined aware medium absorbs and reflects patterns of electromagnetic spatial-energy, or information that crosses the synaptic point of penetration; and reflects and projects that positional state of Being through that penetration point. Once an active energy depleted string is created, a local gravitational field is established around that point; and to that regard the surrounding reflected-matter is affected by that field. This effect reduces and/or eliminates influence of alternate gravitational fields on that surrounding reflected-matter.

ORGANIZED PATTERNS of SPACE-ENERGY 2

In starting from the beginning, we must understand the concept of *Nothing*. Before all things there is no space, no energy, no matter, no time. Before all things there is *no-thing*, so in essence *no-thing* is *nonexistent*. Let me repeat this in a few ways for better understanding.

Here is where physically oriented human beings must slow down and jump off the energy train, and comprehend non-energy related events. The concept of no-thing is understandable; but not comprehensible while a portion of the human mind, as the brain, is full of information. The aware portion of the mind absorbs or accesses information, but this mental state of the mind in and of itself is not-information; not-energy, not-space, not-matter. This mental absorbent nature is only experienced as an absorbent nature, in its capacity to create space-energy and manipulate space-energy. For in essence, the absorbent nature actually absorbs nothing to maintain its conceptual status.

For clearer understanding and repeated here; this no-thing is non-existent, but this miraculous state of non-existent has a mental conceptual place. Search your mind; does the concept of non-existent have a mental place in your mind, a position filled with nothing? You too have the capacity to put nothing in its mental place and position! This mental conceptual place of Nothing, as non-existent, is naturally an absorbent nature to maintain the status as nothing. If the very mental concept of non-existent enters the place of nothing it is no longer nothing. So the place of Nothing in non-existent must be absorbed to eliminate its very place; which cannot exist but cannot be eliminated. So with this paradoxical understanding, the place of non-existent is a never-ending infinitely absorbent and projective place.

To start; be it understood the mental state exists outside of information, outside of space; and be it understood all information is energy; and be it understood all energy exists in space and not outside of space; space is energy, space is time. Be it understood, that beliefs are organized patterns of energy; attitudes and opinions, judgments, sacrifice and abuse, are developed drama through comparisons of organized patterns of energy; but this energy in and of itself, is not aware. As the mental state, an absorbent or Aware place, accesses and reflects energy or information, the aware becomes the rudimentary form of awareness or life; Aware moving through, absorbing and reflecting moving information. It must be understood that ignorance, arrogance and stupidity are informational concepts, and are driven from an overwhelming amount of accessible rigidly held or hoarded information, as beliefs. Belief based information, in and of itself creates nothing but conflict. Believing in a separate God or believing I am God or believing you are God creates nothing but organized informational patterns in separation and at conflict. Knowing all is God outside of space-energy as one body of aware creates everything. This one body of aware creates harmonious conceptual choices, using the unwavering terms for concept creation, to manage the ultimate responsibility for those continuous choices. Aware clouded with belief-held information can seemingly hide from all forms of creation responsibility; and in this regard can ignore maintenance responsibilities, or violate the very creation it is responsible for. This can be seen in the way worship and prayer is conducted. In the worship and prayer of an aware being, that is regarded as separate from our very being, creates nothing and directs onus or responsibility on that separate aware being. When in actuality through responsibility, worship and prayer are the tools used for creation. Worship as an absorbent positional focus, is how a mental concept or ideal is developed; and prayer as continuous mental projection, as responsible creation, as responsible support and as responsible maintenance, for that concept. Worshiping and praying to a Sun God, Mother Earth or an individual gives up creation responsibility, and does not support creation of the Sun, Earth or humanity. Mentally worshiping the balanced concept of the Sun, Earth or humanity into an image or form, and continuous mental prayer for that idealized concept supports the general form of first creation, and continuous responsible maintenance of that creation. Aware that is directly responsible for the creation of galaxies or small solar systems, very rarely dip into the individual timed-space-energy brains of planets that support life, or humanity deeply clouded with space-energy as irresponsible hindsight confusion. But this does not negate the complete responsibility of humanity, to support first creation through love, worship and prayer of first creation concepts. Nor does it negate the responsibility of humanity, while hiding from supporting the conceptual rules that form a life supporting planet. To paraphrase "Stupid is as stupid does", and knowing stupid is hiding in space-energy belief structures from our One Responsible Aware Truth. Aware universe, Aware creator, God universe, God creator, Human universe, Human creator, are appropriately directed responsible terms. God of the universe, God of creation, God of humanity, or the separated singular "God" are inappropriate hindsight terms and create nothing but the concept of completion, of separation, of fear, of control, of power and of size; or individual human irresponsibility as smallness in separation, as judgment and abandonment; as if we are UN-aware. Aware is all encompassing or oneness; size is irrelevant and nothing more than a human created hindsight concept, controlled from within reflected space-energy. The concept of separation comes from Aware riding through space-energy as awareness. This awareness is clouded with information and ignorantly accepts the information as its very existence, but information is created space-energy and not responsible for actual creation or continuous creation of space-energy. Step outside of information and you will know your oneness with Aware creator, Aware universe, and you will know your unwavering responsibility in balanced mental moral choice of continuous creation.

Creation as space-energy takes place for one and only one reason, value fulfillment or worth of Aware as non-existent. Aware is infinite and timeless before and after all space-energy; Aware is the *one mental miracle* as Nothing and space-energy is its *one and only blessing*. To violate the blessing in any manner comes from within the blessing, as awareness in smallness, in the ignorance and arrogance of hindsight confusion and seemingly separations; hiding from the one Aware unwavering responsibility of harmonious projectivity, as rhythm in choice. As human beings our body does not define who we are, nor is the human body owned by our conceptual selves. In actuality there is no true *self*, it is more of a grouping of information your aware spirit is accessing at any given position. The human body, as with all creation, is in a constant state of creation and reflection or re-creation, minute-by-minute, second-by-second, moment-by-moment. We humans as a whole, own the responsibility of continuous support of reflection or re-creation of the human body; but ownership or worship of the human body, in and of itself, implies that its creation is complete; when in actuality every moment of responsible re-creation as reflection slips into the past, and no longer resides in the present moment. In actuality; when the human body can vaguely be considered as done, or created as creation, is in death; for in life, with the synaptic non-reflected spirited energy flow through the bodes nervous system, non-reflected energy refraction through DNA is in a constant state of foresighted reflection; as recreation of its original concept or ideal.

All creation requires responsibly, in actuality the two words are synonymous to some degree. The conceptual choice of Aware as non-existent and actualization of any one position requires continuous acknowledgement. Continuous acknowledgment of a choice, above and beyond any original choice, in actuality is the responsibility of concept development. Continuous responsibility of harmonious projection and absorption organizes non-reflected space-energy into balanced form, and responsible aggressive absorption reflects that form into reflected or re-created space-energy. Actual irresponsibility for creation comes from within creation, comes from within confusion as the space-energy hindsight concepts of size, as in smallness and largeness, weakness and strength, good and bad, develop the logic of separation; as in attitudes and opinions, in awareness as Aware clouded with pools of purposefully segregated information.

Put yourself in a mental place, before space, before energy, before time, from here is where all things started. In this place of no-information the mental concept of non-existent was actualized. This mental state of non-existent has an absorbent nature to balance the truth of the actuality of non-existent. This absorbent mental state must absorb the very concept of nothing, to maintain its concept as non-existent. As the absorbent mental state absorbs the positional concept of nothing, the actual place of non-existent was born. This place of non-existent when considered from within information can confusingly be referred to as black or dark, but the very concept of black or dark refers to space-energy. If this place was referred to in any way it would be referred to as a clear-medium, or clear mental place of the non-existent concept as Aware.

As the absorbent place of non-existent naturally narrows the absorbent characteristic, to one conceptual position of nothing, the position of nothing was born. From this non-existent place and its position as nothing, the spatial-energy barrier between the two concepts is born. Once again the absorbent nature as non-existent must and can only absorb nothing, to maintain the integrity of the non-existent concept. But this non-existent place creates and affects space-energy as an absorbent nature, in its capacity in manipulation of the space-energy universe. This space-energy place between the two concepts is the first form of electromagnetism or light. The space-energy barrier is more of a depletion or vacuum of a non-existent place, developed by the absorption of the positional nothing. This spatial-electromagnetism as light, or the concept of everything, reflects on the skin or shell of the directional or focused absorbent nature, creating the first forms of mental informational absorption and actionable reflection as sight. It must be understood the non-existent place cannot actually absorb space-energy; the conceptual rule of non-existent cannot change, cannot be broken, and so the absorbent, focusing affect on space-energy is recorded and reflected on the electromagnetic barrier of the non-existent concept or place.

So now we have 3 concepts that make up one mind; (1) a clear mental absorbent place, (2) a focal point or position of the mental absorbent place and (3) space-energy field as information or brain. The position of the non-existent place, as Nothing, develops the first foundation as a state of Being; with the space-energy field as information, or recorded testimony as creation, of the first concept or idealized position. This trinity makes up what is referred to as a living spirit, or as recorded; The Father (the one aware creator), The Son (first position created) and The Holy Ghost (testimony as space-energy, or ghostly gray light to first concept, or recorded information about positional creation). This basic wording was developed as a reflecting explanation of first creation, from the limited toolset of ruling hindsight conceptual events, which make up the basis for human social order.

Let's repeat this understanding for more clarity; we must understand what we are and where we are from and where we will always be; the continuing foundation and unwavering responsibility for our very Position, within the one creating body of Love. I say Love here only to make it apparent that the one non-existent absorbent place can be experienced in a few forms. The non-existent place is established as an *absorbent nature*; this absorbent nature has the same fundamental characteristics as the state of aware; or as, aware riding through non-reflected space-energy as *absorbent awareness* or love, as aware laced with energy in motion or emotions; and all these forms create and absorb information with a value fulfilling or blessed nature. The creation of space-energy from nothing as under these circumstances can be considered, or as recorded, Holy, Holy Light or Holy Ghost, as the first form of awareness as life or testimony of creation; as Aware moved through the created non-reflected spatial-energy gestalt. But this absorbent nature from within reflected space-energy, within the hindsight concept of comparison and absorbing desire is referred to as greed. This absorbent nature of greed, as a hindsight concept, is the weakest or worth-less form of creation, in the regard there is nothing actually created. Greed is the shortsighted hind-sighted desire from within individual smallness, for seemingly previously created reflected space-energy or information, within the confusion of comparison and separation, created by the reflected space-energy pool of confusion as to who we actually are. The responsibility of creation is unwavering as a conceptual rule, but with aware deeply entrenched in massive amounts of hoarded information, as ridged self awareness, it can seemingly hide from the rudimentary rule of responsibility; but this developed insanity does not negate the underlining conceptual rule, of complete responsibility for balanced mental moral choice in creation and support.

The creation process is a fundamental movement as Aware; Absorbs, Focuses and Projects, or as foresighted Awareness; Loves, Worships and Prays. The alternate of this creation process involves no actual creation; and is *mentally marked* through horded belief held hindsight confusion and intimidation, but in turn tries to control and protect that which it has given-up or lost responsibility for, as the hindsight concept of Greed; which Compares, Judges and Controls; the *Three Mental Sicknesses* to the creation process.

Aware: Absorb Focus Project
Awareness: Love Worship Prayer
Greed: Compare Judge Control

FUNDAMENTAL MOVEMENT OF AWARE

Creation as a conceptual rule outside of space-energy	Creation or Creativity within space-energy and its reflection as responsibility		Creativity within the space-energy reflection through comparison as shortsighted to creation responsibility
Absorption – Aware		Love – compassion; absorbent nature laced with created space-energy as absorbent awareness	Greed – self absorption through separation as smallness through comparison
Focus – directional absorption		Worship – harmonious choice in development of individual concepts as positions	Attitude – selection through opinion of creation comparison as judgment
Projection – directional compaction		Prayer, Song, Chanting – supporting continuous harmonious thought through movement as emotion	Control – creation-less irresponsibility, as aggressive supervision of separations in space-energy to experience largeness within smallness

It must be understood that the basis for an *individual's worth* is basically the gathering of individual energies as worth. An individual's energy is the worth of individual creation and/or creativity; whereas an individual's power, be it ill-gotten or freely given, is individual control of taken and/or given individual energy and can be mistaken for worth. Worth is created, Power is obtained. Worth is infinite, for this is the value fulfillment of creator; Power is size oriented, timely or temporary and controlling. But individual power that responsibly manages multiple individual energies, as group-worth, is responsible value-fulfillment and everlasting for all.

It must also be understood, that individual conceptual rules cannot be altered; new concepts can take form from an original concept, but these new concepts do not and cannot alter the original concept. We must understand concepts are mentally developed and are unwavering rules; in and of themselves. If a concept could be altered, the rule would be broken and the original concept would no longer exist. How can a concept no longer exist in truth, if its actualization is recorded in space-energy or behind us; it can only be forgotten in the confusion of new or seemingly new concepts, as aware drops its positional focus.

So the rule of a given concept cannot be destroyed, it can only be left behind as forgotten. So the rule of the non-existent concept and its place cannot be destroyed, and the rule of its position as nothing is unwavering. The true rule of non-existent must absorb its very concept as nothing, for the concept of non-existent to maintain its place. Seeing this paradox, as the great contradiction of non-existent that cannot exist but cannot be altered or destroyed; it in actuality is an absorbent nature, to maintain the void it is. This impossibility is the actual argument or position; held by the one ruling non-existent concept. We must understand that punctuation within this very document are ruling positional concepts, mentally developed before their very form or outward appearance and if the form is not recognized, the concept in and of itself is unwavering. As for me, the writer of this document, my understanding of punctuation concepts could be in question and rightfully so. As the very concept of non-existent, in and of itself is in a constant state of question as nothing and rightfully so.

To repeat; these basic understandings originate outside or before creation of space-energy-time, but are recorded in and on the shell of space-energy-time. From within the non-existent or mentally absorbent place, it must be understood that the one absorbent action has a corresponding reaction as projection; both absorption and projection can be considered as a magnetic nature, when experienced from within spatial-energy. As with the absorbent nature, the projective nature only has an effect on the space-energy field and not within the non-existent place. As an overall absorbent nature (non-focused) the reactive projection is centralized or within this absorbent place with no affect. As with a focused or directionalized absorbent nature, the reactive projection is opposite to the focused absorption with no affect. Let it also be understood this reactive projection characteristic can be directive, as the absorbent nature can be focused. So absorption can be mentally focused and projection can be mentally directed.

Humans can experience all these affects, within the space-energy field, by simply focusing on an object without thought and absorb that information. The information or space-energy is transferred and absorption takes place between the sight related synapses in the brain, and to some degree, and in a more general manner, in the gap between the right and left halves of the brain. This gap between the brain halves acts like a large synapse and is mistakenly called the third eye, but in actuality when used for absorption would be one addition to all the sight related synapses in total. In this situation the mental projection is behind each synapse absorption point. It must be understood the basis for thought, is a mental state of absorption, reflection and projection; through synaptic absorption and reflection of information, and action as synaptic projection on that reflected information. This is all experienced as humanity absorbs the space-energy around them, and projects their thoughts out into that same space-energy, and in some cases this projection results in individual or group action. It also must be understood there are groupings of synapses dedicated for absorption and groupings dedicated for projection. The absorption synapses are segregated to manage much more than sight related space-energy absorption (i.e. the energy sent through the nervous system relating to the sense of physical contact), and the projection synapses are segregated to manage much more than individual thought projection (i.e. the instant projective reaction created by the sense of physical contact). In actuality; language is a rapid mental process of absorption and projection that creates non-reflected space-energy in unison with the rhythmic organization of that space-energy into structured form; as a form of energy in motion or emotions. This is a very basic understanding, but difficult when absorbed through the confusion of specialized reflected space-energy informational structures. After all, the human body is a space-energy gestalt of accessible information; no different than a solar system is a space-energy gestalt of accessible information. This information may vary, but does not exist outside of the space-energy field.

Now we must understand there are two basic forms of space-energy, non-reflected and reflected; or as non-physical energy and physical energy. Non-reflected energy is mentally created and mentally manipulated energy, and reflected energy is non-reflected energy in reflection. It must also be understood; once reflected this physical energy can be manipulated, but it is not mentally created or mentally manipulated in the same sense non-reflected energy is. It must also be understood that reflected energy can be re-reflected in varying degrees, creating more precise or rigid forms of physical energy. It must also be understood; not all non-reflected energy is reflected, and there is a limit on the re-reflection of reflected energy. Again it must be understood; all forms of energy can be reflected-on as recorded by the non-existent or mental medium, but not necessarily physically reflected. Reflected energy is more of an altered re-creation of non-reflected energy, which is reflected off the non-existent mental concept. The amount of reflection, as re-created, is in direct proportion to the intensity or effort of focused absorption; but reflected-on energy, is energy that is experienced as recorded but not focused on and absorbed to the degree of re-creation. Reflected energy is what the majority of humanity can readily identify with. This physical energy is the only form of energy the human eye, made up of a short bandwidth of physical energy, can react-to. It also must be understood there are forms of reflected energy that is not detectable by the human eye. It also must be understood that instrumentations created with physical energy, of a specific bandwidth and decay rate, cannot detect non-reflected energy or reflected energy of differing bandwidth and/or decay rates. As in the misguided saying heaven has many rooms, there are many bandwidths and decay rates of physically reflected space-energy. Even though the human body we identify with is a combination of both forms of energy; the non-reflected energy portion of the human body is dismissed through tightly held intellectual or material belief structures. Rigid information, of a physical or material nature, will not simply be experienced in other physical energy bandwidths, and/or a non-reflected space-energy field; or as recorded, pass through the proverbial *eye-of-the-needle*.

It must also be understood that the focused aware medium can reflect-on as react-to or see all forms of space-energy, but can interact with mentally created non-reflected space-energy to a much greater degree. This greater degree of interaction is referred to as awareness, aware interlaced with non-reflected space-energy in motion or emotions. Humans can readily identify with this non-reflected space-energy in motion through personal, social or global experiences, through individual awareness as aware feelings. The amount of one's actual experiences, of non-reflected energy in motion, is determined by one's individual openness to the one aware medium; and to some degree detachment from the narrow physical energy bandwidth, that makes up the human body we temporarily occupy. It must be understood that the human body takes mental form in non-reflected space-energy through mental manipulation, and up through the physical or reflected space-energy bandwidths through mentally developed DNA; before actualization into the most physically reflected environments. These other space-energy bandwidths are visible to varying degrees during the differing stages of sleep. This visibility of alternate space-energy fields is mainly made possible through individual correlating body's visual organs, in altering space-energy bandwidths and decay rates, that report back to the same body of aware; as an individual spirit is associated with through keyed DNA interaction and absorbing synapses. But it also must be understood in some cases; nonrelated space-energy bandwidths are visible during waking hours, through the one main synapse between the two halves of the human brain. This one synapse does not use the narrow physical bandwidth making up the human eye, but directly absorbs information in a much broader sense. But the human brain will react to this information through the same channels, as if it were viewed through the physical eyes. This also holds true for sound, which is absorbed through the main synapse between the two halves of the human brain. This activity once again is generally dismissed through tightly held intellectual or material belief structures, to manage a sense of stability in massive group smallness as insecurity. A conceptual belief or disbelief structure creates nothing; and are generally very tightly held small energy pools of hoarded information, at conflict with seemingly large or un-comprehendible information, but allow the confused ego or self a sense of awareness security.

ORGANIZED PATTERNS of SPACE-ENERGY 3

It must be understood the multiple physical bandwidths of space-energy are made up of differing levels of reflected energy. Within these differing physical fields of reflection; the ruling effect of absorption and projection (i.e. gravity) react differently on less physical reflected space-energy than highly reflected space-energy. These differing affects are confusingly experienced during what humanity refers to as dream experiences; where the human body has differing capabilities, and/or experiences that seem extremely foreign to our waking reality. It must be understood that these dream representations of our waking human body are truly individual forms of ourselves, in different physically reflected realities and seemingly operate independently, but never the less operate from the same never sleeping spirit, and individual, group or social personality pools of information.

For clarity; it must be understood there are three forms of space-energy creation:

- First creation, as non-reflected space-energy…
- Second creation, as reflected and re-reflected space-energy, as re-creation…
- Third creation as creativity, as organized manipulation of first and second creation…

It must also be understood the human race as we know it, mainly identifies with Third creation, as in the arts, as design, engineering, science, literature, etc. These forms of creation are more associated with creativity or creative manipulation and study of space-energy, and not actual creation or re-creation. It must be understood the physical human body we generally relate to is of the Third form of creation.

To get mentally concept created organized patterns of non-reflected energy (i.e. a mustard seed), from the mentally created non-reflected space-energy field, to reflected space-energy requires a great deal of effort of mentally focused absorption and projection; to reflect or re-create that complex pattern as physical energy. The more rigid the physical reflection in heavily populated reflected space-energy fields, the greater degree of effort of mentally focused absorption and projection is required. But with the mentally developed and mentally manipulated elemental concepts of DNA, this reflection process requires far less mental effort; in non-reflected energy refraction to reflection. This DNA refraction process does not directly reflect non-reflected energy into physically reflected energy or atoms. In actuality and to altering degrees; the flow of non-reflected energy through the elements that make up the DNA refraction process; alters or changes pre-created physical energy or chemical elements into alternate chemical elements, by altering the number of reflected protons of a given atom in a given element. Simply put this process changes the chemical elements, into alternate family elements, within the periodic table. This altering of physical elements, into alternate elements, can be experienced when a directed non-reflected energy flow passes through that physical element; as in the extremely misguided understanding and practice of what would be referred to as alchemy. This DNA non-reflected energy refraction to reflection can actually create rudimentary elemental compounds; when the non-reflected energy flow through a DNA source is sufficient, a burst of physical material will be created. As for a mechanism of this order, this capability is unexplainable in a belief oriented social order; and simply referred to as a miracle or manna from heaven. As for a DNA source; uniquely cut crystalline materials can mimic keyed DNA non-reflected energy reflection.

As is apparent to all living organisms on our planet and to varying degrees to most human beings, there is a larger force at hand in the creation of the universe, then just the very narrow physical energy bandwidth we can readily identify with. This lack of understanding, as confusion for the human race, is generally developed through an awareness bubble, of strict organized patterns of information, required to individualize the aware medium, to react to activity within a sphere of influence, or physical energy bandwidth. This bubble of personalized information or awareness, can be referred to as the human ego or self, and is more of a security blanket that expands to societies as a whole, to manage some form of social activity and order. This bubble of information is used throughout the differing physical energy bandwidths; but it must be understood these altering physical bodies, within altering energy bandwidths and decay rates, is merely information that is not generally available to our waking selves. Once again; it must be understood the aware medium and individual positions within that medium, do not and cannot actually absorb information; so the information regarding the differing physical realities is not stored in the aware medium, for retrieval in alternate physical bandwidths. This alternate information, which reacts in differing ways as stored in alternate bandwidths of reflected energy, is and can be glimpsed or correlated in one's current reality; if that alternate information is absorbed, and re-reflected as recorded without sacrificing the integrity of an individual ego or energy bubble. If this information bubble or ego is not recognized, and managed with some flexibility, it will be degraded with alternate information, and individuality will be compromised.

The actual flow of mentally created non-reflected energy, into the reflected environments, is laced with the aware medium; so in a sense this is a flow of awareness into physical or reflected environments. Again; awareness is aware riding on mentally created non-reflected space-energy, as information. But awareness does not reflect-on that information like the positionally focused aware medium. Actual reflection, on specific information, takes place at individual synapses in the human body. The synapse gap is actually a pinpointed tear or hole in the space-energy field, back to the non-existent concept, or positionally focused aware medium. This aware medium is not awareness; there is no space-energy as information in the non-existent concept. It must be understood; the aware medium has no memory, the only way a memory works is for aware to access information. This information can be newly created, or stored information as memories. This is where the human intellect falls apart when dealing with the basis of who we are. In general; we relate to our awareness and not our aware. Awareness can be referred to as emotional aware, or aware in space-energy motion, but this form of aware as awareness accesses seemingly previous created information; and does not generally create non-reflected space-energy. It is more like Aware using space-energy, than actually creating it; this awareness is a good place for humans, entities, souls and over-souls to hide from creation responsibilities; and remain oblivious as to what is actually going on. It must be understood the human body is a grouping or gestalt of space-energy, no different than entities are groupings or gestalts of space-energy; whereas souls and over-souls are more of the timeless aware or mental component, surrounding, creating and accessing a given position and/or positions; all of which are accessible by the Trinity or living spirit. This basic grouping is more specific to individual reflected energy bandwidths, and as the space-energy groupings expand and encompass more information awareness becomes more general; but it must be understood that Aware, as a living spirit, within any informational order, can lose the sense of complete and total responsibility as creator.

It also must be understood; this space-energy free gap, we refer to as a synapse between two nerve endings, can be recreated in a larger form. The DNA within the nerve endings creates a refracted non-reflected energy reverberation, within the gap and creates a tear in the space-energy-time continuum. Once again; this capability of tearing the space-energy-time continuum, if created outside the human brain, is unexplainable in a belief oriented social order; and simply referred to as a container or box that performs miracles. As for a mechanism of this order, the basis for the DNA saltwater solution, as a non-reflected energy refraction source, could literally be explained away to the shortsighted, as the proverbial blood of a sacrificial lamb. The mechanism of this order was rightfully shielded, or referred to as being guarded, due to the fatal affects it had on rigid hindsight belief oriented energy gestalts as physical human personalities; through the massive flow of non-reflected energy moving into reflection, this mechanism established as balanced emotions and reflected as physical electromagnetic energy; then simply discarded through lack of understanding and fear. This non-reflected energy flow is misguidedly experienced as enlightenment, nirvana or referred to as chi energy, which can have profound effects on hindsight concepts and the mentally irresponsible or undisciplined. A mechanism of this order would give off a non-reflected energy white-light or hallow glow, with a low hum produced by the refracted energy clash as a reverberation; then as the non-reflected electrons move to reflection a steady electrical flow to earth ground can be established and/or alternating electron flow between the so-called mechanical nerve endings. If this non-reflected energy flow was sufficient, and then passed through an individual's nervous system, a spontaneous combustion effect would be experienced; or as misunderstood very basic wording and simply put, one would burst into flames. Whereas; a managed moderate flow of non-reflected energy through the biological electrical system, as the nervous system, will improve the DNA reflection process supporting health and longevity.

It must be understood and repeated here for more clarity; the aware medium absorbs no space-energy, but has an absorbent affect on space-energy. Space-energy that is absorbed by the aware medium is reflected, to varying degrees off the focused aware medium, and is reacted-on by the aware medium; as in re-creation of space-energy as recorded action. It also must be understood and repeated here for more clarity; the aware medium has no space-energy, and has no capacity to record or store information that was reacted-on; information is only stored as space-energy in space-energy. The aware medium can only recall information that was reacted-on by re-absorbing and re-reflecting-on that recorded information, as recalling a memory. So it must be understood this lack of ability of the aware mental medium to record information, is what makes it so difficult for humanity to comprehend, remember and accept their individual truth. It must be understood that recorded or memory information in space-energy, is only recorded in the space-energy bandwidth that information was reacted-on. So information from other space-energy bandwidths is not readily available to our waking selves, but can be experienced when an individual's aware focus is expanded, mainly in a sleeping or meditative state, and remembered if that information is recorded in one's current space-energy bandwidth. But generally this information is not focused on as amplified, and simply left behind as forgotten.

It must be understood and repeated here for more clarity; awareness is aware accessing information and falsely considered as one's actual existence or position. This is why it is so simple to remain small and separate from the responsibility of creator, by identifying ourselves as space-energy creation and not creator. Once again all space-energy is creation and Aware, outside of space-energy, is creator and mental manipulator of space-energy. As in reading a book; our aware rides through the words as awareness, and we can readily identify or get lost in that information we identify with. We may actually go as far as taking ownership of this book of information, and place it in our personal energy gestalt or library. Aware riding through or getting lost in our human body of space-energy, as awareness works in much the same manner, but in this instance we did buy the book and place it in our personal library. It must also be understood the very concept of demons, devils and the like are nothing more than individual ignorance and arrogance, as awareness in smallness seeking power and control, negating the very responsibility of creator. To except one's position and responsibilities as part of the one Aware medium, one must step out of the space-energy box. Aware is infinite, and the sense of urgency has no bearing or reference; but as with awareness, all space-energy is in movement and time oriented.

It must also be understood and repeated here for more clarity; each and every human body is developed from a first concept, from an overall humanity first conceptual pool; it's not who you are, it's more how your body is created within the creation process. The perfect mental image or ideal of humanity glows from within non-reflected energy, and out through individual DNA as re-creation. This first concept or worship is supported through continuous projection or prayer of that concept, through non-reflected space-energy manipulation, up through the reflected version you identify with during your moment-to-moment of re-creation. It must also be understood this first human body concept is perfect in and of itself, but is degraded to varying degrees from what seems internal and external informational influences. All degradation of original mental concepts and re-creation comes from distraction, through multiple concept reflection and choice, that can lead to comparisonism; the proverbial forbidden fruit. This reflection through comparisonism can be considered the first mental sickness to creation if allowed to manifest into attitude and opinion. Attitude and opinion is derived from the degradation or worthlessness of some or all of the compared first creation concepts. If attitude and opinion are unchecked, this mental sickness will develop judgment and superiority, sacrifice and abuse of first creation, as aware blinded within space-energy from all creation responsibility. This mental sickness to creation deepens and unfolds as drama, through opinion of opinions, developing attitudes of an attitude and judgment of judgments, etc. To make a clean, clear, pure individual choice between concepts, which fit within the conceptual ruling pool you are part of, leaving other concepts behind as forgotten; is much different than making a choice through an opinionated judgmental degradation, of the non-chosen compared concepts. This form of choice through opinion and judgment actually leaves the chosen concept degraded in varying degrees, as it leaves all creation degraded in varying degrees. It must be understood that a conceptual ruling pool is a pool of non-conflicting or balanced concepts, which make up a particular grouping or species. It must also be understood; in stepping outside the human non-conflicting ruling conceptual pool, with a given choice can degrade or pollute that overall pool to varying degrees; similar to polluting the earth with overpopulation through unchecked judgmental belief structures, allowing the release of personal and social responsibility of life supporting planetary concepts.

To repeat once again for clarity; Aware creates all space-energy from a timeless place through choice, continuous space-energy creation requires continuous choice; ALL choice requires constant effort as responsibility. Dropping effort drops responsibility and creation as space-energy will decay or dissipate. Created space-energy is not aware; but Aware riding through space-energy is Aware in value-fulfillment of creation, as awareness. Aware does not intentionally make choices that are in direct conflict with alternate choices, there is simply no value-fulfillment or worth in it. It must be understood that making a conceptual choice, to create or manipulate space-energy is creator; but making a choice between compared created space-energy is not creator, but both require complete responsibility for that effort or action. Portions of the mental state of aware lost in space-energy-time as awareness, experiences a sense of separation or smallness; this is not actuality, but more of Aware experiencing its creation; and more often than not for the human race, a good hiding place from its true nature as responsible creator. A choice that is in conflict with an original and continuous harmonious choices; that create, or support life supporting solar systems and planets, are not bad or good; they are just destructive, and have no worth within a balanced massive effort driven, chosen and re-created structure. Whereas in actuality, there is no creation; seeing creation is in a constant state of fulfillment, moment-by-moment from a non-time oriented place. Creation is a shortsighted human concept, as Aware picks through the moments of life, implies creation is completed or finished; when in actuality, creation is in a constant state of space-energy combined amplification and decay as time. Human awareness that is strictly tied to or not willing to release information, as a given belief, that is in conflict with creator responsibilities, as disassociated information, as a conflicting awareness personality; when deceased will experience a sense of smallness, loneliness and despair; or as cast-out into a seemingly abyss when that information decays as dissipating; or as that information is left behind as forgotten or dead to further development by the overall Aware creator responsibility. To repeat; ridged materialistic human personality belief structures that cannot and will not move to a conceptually altered as foreign space-energy bandwidth, will be simply left behind as forgotten or dead to any further personal Aware interaction. Aware as awareness in this situation can eventually release this conflictive information, through an alternate focal choice, or until such information has dissipated through a lack of amplifying focal support as boredom.

To repeat; the aware place can be expressed as unmentionable or unspeakable, and within the human oriented information or belief structure, this is confused with being so holy that humans are not worthy of speaking the name. When in actuality; there is simply no space-energy wording that can accurately describe no-thing, as an all encompassing, time-less, size-less, absorbent, focused, projective, non-space-energy, non-existent, conceptual aware place. All space-energy-time is in a constant state of creation, through the mental choice of focus-absorption-projection as creator. All life is space-energy laced with the one non-space-energy Aware medium; believing or disbelieving in this is irrelevant, beliefs are nothing more than pools or gestalts of specialized created information as space-energy, laced with confused Aware that may take up no personal responsibility as creator. Creation is not created as done, but in a constant state of being created, as a flow of non-reflected energy, and in a constant state of re-creation as reflection, as in the ever present moment. As true with responsibility; responsibility is not required as done, but in the state of continuously being responsible for the ever present moment of balanced choice and reflection.

It must also be understood that all predictions, prophecies and history are neither true nor false. For all past and future events are in a constant state of being created, moment-by-moment, within their momentary time frame. To that regard; to prove any post event as true or false is just an act in futility, and takes no responsibility for the current moment of creation, and actually creates nothing in the present moment; except conceptual comparisonism as abuse, as one truth in belief is excepted and the other at conflict. For after all; the very notion of the non-existent concept is in a constant state of contradiction, as all concepts past and future are in a constant state of fluctuation, as contradiction, as managed in the responsible mental balance. We as a people must live in the personal moment of pure and constant responsibility in blessing from our very own Aware; for this, our reality is the one and only blessing as space-energy. But if humanity needs to hold onto irresponsible horrifying probabilities, as actual events, to keep their moral balance as creator; then they do serve a fulfilling responsible purpose. Or to that regard; if humanity needs to hold onto blessed probabilities, as actual events, to keep their moral balance as creator; then once again they do serve a fulfilling responsible purpose. But societies as a whole must be wary of drowning in information and comparisonism, generating massive amounts of unnecessary information as confusion, which can and will become un-processable. We as unmitigated mentally responsible creator, must not sway from this truth with an over abundance of recorded information. The ability for individuals to access and process large amounts of information can be seen as a form of intelligence, but if the responsibility to truth is neglected the capacity as aware creator is lost. It also must be noted; past and future probabilities that have profound influence on our current moment of creation, can and will ripple out and affect multiple momentarily timed corporal realities; unless recorded as guide posts for understanding and retrieval. This recorded information can be considered as more of a heads-up guide, for individuals or societies, and is generally retrieved through historical recording and sites or lucid future recordings, as memories from dreams. After all; all space-energy as the blessing, as information, is created by Aware and retrieved by Aware in the journey to fulfillment as nothing; and as universal humanity, throughout non-reflected and reflected environments, as the pinnacle, as most personal, psychologically cognizant, thoughtful sentient, self awareness, form of that Aware.

ORGANIZED PATTERNS of SPACE-ENERGY 4

Unmitigated responsibility for conceptual balance; without unmitigated responsibility for balance in choice in the non-reflected environment, the reflected or physical reality we know and love would never get off the ground; so to speak. This non-reflected environment is created by Aware, but is not Aware in and of itself; but more as managed by the deep profound responsibility, of balance by Aware as awareness. Let it be very clear; personalities as informational gestalts that reside in this non-reflected unmitigated balanced environment, that reflect as re-create physical realities, are without conflict in choice. Tumultuous personalities will simply dissipate in this environment; but the individual pure trinity or spirit as position in being, without conflicting information, will be right at home. Pure balanced personalities from this realm are misguidedly referred to as angels or deities, but in actuality are just highly evolved in balanced worship and prayerful choice. But again; a personality gestalt of this nature is not Aware in and of it-self; but more accessed by and created by the Aware living Trinity or spirit, through unmitigated responsibility in conceptual choice. Destructive information will not and cannot move to this non-reflected space-energy realm, for there is no foundation or bed of information to leach onto. All mental concepts are created in the ever present moment, so in this non-reflected realm these concepts are in a constant state of creation or dissipation; so in essence there is truly nothing regarded as permanent, to compare or have an attitude about. But with the reflected or re-created energy delay from creation, there is a false sense of permanency; as a form of hindsight, that creates a bed or foundation for destructive comparisonism concepts; as a worthless sense of controlling, all be it fleeting, insanity. But again; this generated insanity is immediately left behind, when the pure Trinity or living spirit moves to the non-reflected realm; as is confusingly labeled heaven. A personality gestalt can leave a re-created environment with its worth, but its power and wealth will be left behind. In leaving destructive belief based hindsight created information behind, as reference to the misguided concept of judgment or atonement for ones sins for this worthless behavior; but to that regard, this information simply will not, and cannot exist in the non-reflected moment-to-moment first creation conceptual environment. But again, this is not Aware; it is Aware making balanced choices that creates and organizes this non-reflected space-energy, and unmitigatingly balancing these non-reflected space-energy patterns as awareness. But as the reflected space-energy pool moves through the stages of re-re-reflection as re-re-creation, this momentary sense of delay or hindsight is elongated; developing the notion of supervision and control, allowing for comparisonism to flourish into self degrading reflected energy concepts, degrading first creation re-creation fulfillment. For in truth; violation of a concept, which is in the never-ending process of becoming, cannot take place except through hindsight of creation; through the reflection delay. All concept creation is a mental process; mentally created hindsight concepts will become re-created in the reflected environment, if sufficiently amplified through prolonged individual or group mental awareness focus, and if destructive, will be re-created in the rich-bed of re-creation as disease or cancer to first creation reflection. Once again; this is how humanity as a whole hides from their lack of mindful mental responsibility, through the numbing delay of re-creation. It also must be understood; hindsight mental concept creation is neither bad nor good, and is required for humanity to survive in the re-created environment; as seen in the simple hindsight concept that drives continuous mobility in evolution; as food foraging and consumption. Whereas; hunger and the heart beating are more along the lines of first creation concepts, to maintain mobility in the reflected environment, directed from within a balanced non-reflected environment, and breathing is a mixture or balance of both. Whereas the first creation concepts of *love, worship and prayer* of an ideal into form, into reflection as the blessing, would be supported in the hindsight concepts as continuous prayer as *perfect health, perfect physique, perfect form* in regards to the first creation ideal, but not in regards to dramatized hindsight comparisonism. Or as in the every day and most effective hindsight concepts of good morning, good afternoon, good day, good evening and good night or as more personal, good dog, good boy, good girl, etc. And whereas the old adage of hindsight being 20/20 actually creates nothing, in the moment-to-moment of creation; but this hindsight information is recorded and can and does affect future mental projections and creates a bed of intuition for future moments of choice. To be very clear, the *creator* concept is a first or foresight concept and the *creation* concept is hindsight to creator concept; whereas the creator concept implies responsibility at any given moment and the creation concept releases responsibility into the illusionary past. It is critical for a healthy society, planet and solar system for all forms of humanity to replace comparisonism power driven concepts with clean, clear, pure observation, with a supportive loving, worshiping and prayerful choice; without insecure and controlling reflection on what another is doing.

Now it must be understood and repeated here; deities or angels are personality gestalts or structures made of constantly balanced first creation concepts, where second creation concepts as hindsight concepts do not and cannot exist; in repeat, there is simply no foundation for them to leach onto. But again this is not who they are, for they are aware as we are. The living spirits of these personalities are riding through, as awareness, first creation concepts that they create. But in this realm as aware spirits they are completely mindful of this self created value fulfillment or blessed illusionary worth; for as Aware as non-existent is nothing, regardless of what space-energy you are creating and/or accessing. As with all living spirits; the created space-energy event-horizon or Holy Ghost, is only the recorded testimony to Aware's position and movement and not actually Aware. The spirit or state of being naturally equates its very being with that that has been created or what it creates as space-energy, for this is the only comprehendible reality; but in actuality Aware as the clear-medium, as a mental medium, is not readily actionably comprehendible as an absorbent nature. All space-energy is accessible and acceptable as the blessing, but Aware within the living spirit is not readily comprehendible as an un-known absorbent miracle. Living spirits created within the clear-medium are more acknowledgeable as creator and responsibility than living spirits or positions created within space-energy, within reality, within the blessing; due to the nature of space-energy confusion, as to what they actually are. To this regard; individual personality gestalts (human beings) created from a balanced mental profile, as recorded within the physically reflected brain; that we falsely identify with as to who we actually are, can and will move into the non-reflected realm with their spirit, or as misguidedly referred to as going to heaven or eternal life.

As within the *non-reflected, or foresighted, first creation concept of male and female*, there is no true separation; but to that regard, the male portion in general creates and manages the vast realm of balanced first creation concepts, but has a tendency to drift through curiosity; whereas the female portion has quick access to this vast realm of balanced first creation concepts, for precise support as massive worth, up through fulfillment, but also has a tendency to drift through choice. In general these specific peculiarities go hand in hand. Aware as creator of first creation concepts, would not and could not move into re-creation if this union were negated or neglected, or put in another way; the male portion is more about concept worship (absorbent focus) of first creation into form, and female is more about prayer (projective e-motion) of that worship into continuous form up through re-creation; or as quite literally, from within the non-reflected foresighted environment, female giving prayerful birth to those male worshipped ideals. Or as in the planetary concept of size as established in re-creation hindsight; as mother-earth, when in actuality the female concept is infinite as, mother-solar system, mother-galaxy, mother-universe, and whereas the male concept is infinite as, father creator; whereas in actuality and outside of smallness, outside of creation; both operate as one in the one aware place. As both worshiping (male) and prayerful (female) personalities in the non-reflected realm; must evoke a massive responsibility as creator to bring forth the fulfillment (blessing) of first creation concepts into non-reflection up through reflection reality. But in a re-creation gender-based male female orientation, these peculiar characteristics can and do drift; seeing there is no true separation in the basic foundation as Aware, as creator. Whereas the plant and animal world are more living awareness within the blessing, as creation; and whereas the human race has a footing in both, as creation and creator; with unmitigated responsibility (morality), in managing choice to span both realms as mental creation and creator; with the capacity to move individual information, as a personality, to one's endless living spirit. These re-created gender-oriented male and female concepts fully complement one another, until they fall into the trap of re-creation hindsight concepts such as, separation, comparisonism, competition and control. *We are one _universal_ people*; pinpointed smallness in re-creation, in the false sense of permanent hindsight concepts, as seemingly testimonial recorded information, is *fully _immaterial_* to _universal_ Aware or creator within us.

We are in Aware, as Aware is in us; we *are* Aware!

Mental projection can be broken down into three very basic mental tools or understanding as mental sight: foresighted, shortsighted, and hind-sighted.

- Foresighted is associated with the aware or mental component of the living spirit.
- Shortsighted is associated with non-reflected space energy as the awareness portion of a living spirit.
- Hind-sighted is associated with the awareness portion of the living spirit in reflected, as re-created, space energy as in physical realities.

In physical realities all three forms of mental sight are in play, which can and do rapidly overlap.

Foresightedness is associated in creating a focus as choices, as freshly identified ideas or positions, as concepts or as actionable intelligence.

Shortsightedness is more associated with managing specifically identified groupings, as organized patterns of information in non-reflected through physically reflected environments. Shortsightedness is also used to manage belief-based information as social order, to family, to global history, which can be of stabilizing benefit or result in rigid conflict.

Hind-sightedness is more to manage the reflection delay as re-creation as physical environments; as in looking back at reflected as recorded information as the physical environment, as events to make foresighted choices that can impact the current moment.

Shortsightedness overlapped with hind-sightedness can and will generate attitudes and opinions in reference to comparisonism of those shortsighted groupings of information.

The shortsighted mental sight bubble can vary regarding rigidity and is associated with being emotionally thin-skinned or thick-skinned; that is managed with individual or group self-worth and security.

To repeat for more clarity; Aware directs its focus as created positions and those positions are supported through constant repetitive projection, as responsibility, and as support for continuous creation of an individual position. These created positions are the only form of value or worth, as a developed fulfillment, as non-reflected space-energy. As an original focus or choice develops in harmoniously overlapped choices, and complexity, and eventually reflection, continuous creation responsibility is magnified and critical for a stable and balanced re-creation environment. The only worth Aware as nothing has is responsibility; in continuous harmonious movement as *love, worship and prayer*, as the *blessing and creation fulfillment*. Without the creation of *space-energy*, Aware is truly nothing as non-existent. Take ownership of creation worth through responsibility; take ownership of creation-less worthlessness through hindsight judgment.

Loving worship of reality and humanity into form and prayer-full projection of that form as support into reflection…

As hindsight of creator takes subtle control for evolutionary mobility; the harmonious sub-consciousness mental responsibility turns to conflictive sub-conscious neglect and decay, through confusion.

For clarity, as seen from within complete responsibility as creator; when born into a reflected or re-created environment, one can be considered as being born into delayed confusion. These delayed re-creation confusions, within the illusionary sense of permanence, which generates hindsight ideals for evolutionary mobility; are unique, that can and do become non-supportive and irresponsible; to those ideals generated in the moment-to-moment non-reflected first creation environment. This can be seen from a very limiting toolset of basic wording and understanding, within a reflecting explanation, as being born into sin; but in actuality the reference is being born into irresponsibility to creator as Aware within us. This misunderstood information can and is used in the hindsight ideal of original sin, and when coupled with the greed filled hindsight ideal of compare, judge and control, to manipulate individual worth, generates the most degrading hindsight ideal to creator as condemnation. This is all done to gain the illusionary sense of falsely legitimized control and power. These building blocks of degrading hindsight ideals are the most worth-less and destructive to non-reflected first creation concepts moving to reflection in the re-created environment, and are seen by Aware within us as a sickening mental illness to responsible creator. When in actuality this controlling desire is a hindsight concept to stop the unjust degradation to first creations balance or rulings; that can and does get out of hand with weakness and horded information that is strengthened by greed filled power. Responsibility as creator is complete, regardless of an individual's capacity to manage that complete mental responsibility. This limited capacity to instant gratification to fulfillment in re-creation, as creator in responsibility, within the massive numbing confusion delay of reflection; develops a form of complacency, incurring a mental laziness in movement for creator responsibility; including one's very health and well being for moral balance in their own humanity. Individual and/or group worth is the value-fulfillment, which you can create and/or support through individual and/or group capacity for that responsibility.

To make it very clear, so we are fully AWARE; the aware medium, is the clear-medium, is a timeless place. The living spirit is a position within the clear-medium, with a non-reflected energy barrier around that position... The living spirit is a trinity, made up of, aware, non-reflected energy and position; whereas, aware and position, are timeless, as in NO space-energy; the non-reflected energy portion, is a non-reflected energy barrier, around its position as non-reflected electromagnetism. This positional non-reflected electromagnetism, as positional self awareness, is timely, as in fluctuating between its amplification and decay rate, managed by the aware positional focus surrounding it. This energy barrier is in a constant state of creation and support by the Aware positional focus, and will dissipate if that positional focus is dropped; whereas the Aware or clear-medium will lose its sense of timely positional being, but the actual Aware place remains infinite. This timeless Aware and timely awareness living spirit creates, supports and accesses, to varying degrees, the timed space-energy universe through directional focus, absorbent and projection; that can and does get very specific, as pinpointed, accessing the reflected space-energy universe through the synaptic gap, that is keyed to a specific DNA signature. This living spirit can access non-reflected and reflected environments, and can simultaneously access multiple bandwidths of reflected environments through keyed DNA signatures, within those environments, as we relate to in dream experiences. All these space-energy environments are constantly being created and supported through the all encompassing size-less, time-less Aware medium; in union, with the ever moving universal non-reflected energy positional awareness mediums. Whereas in the plant and animal world; expanded living spirits will access multiple keyed DNA signatures, within one reflected energy bandwidth; whereas individuality is expanded within that species, according to their unique characteristics; as in energy gestalt, or energy profile, that that informational grouping is related to. It must also be understood; with tuning capabilities and structural modifications to the presented concepts; shifting between universal locations and/or alternate reflected timed bandwidths is possible by appropriately responsibly balanced mental personality profiles. It also must be understood; an active synapse develops a clear-medium point or hole in a reflected space-energy field; but this is not positional focused aware, as in a living spirit.

Humanity as we know it is well aware of experiences that have been unexplainable from our small reference of information, but nevertheless are actualized and can be re-created in the concepts presented here. It must also be understood; the presented information is intended as *Actionable Intelligence Only* or *left behind as forgotten,* and *not intended to degrade, support and/or be use as an informational belief structure of any kind.*

Figures 13-15 depict perspective views showing event chain from lettered a-l: conception 1, conceptual position 2, positional energy barrier 3, dashed arrow as conception focal direction 4, solid arrow as conception projection direction 5;

Figures 13 positional energy barrier amplification b3, focal and projection circumferencing directional shift c4 & c5, cylindrical energy barrier d3, projection directional shift e5, energy barrier collapse e3-h3, shelled ringed thermal reaction 6;

Figure 14 projection directional shift i5, amplified positional energy barrier i3, projection directional shift j5, multiple shelled ringed thermal reactions j6, collapsing energy barrier j3 & k3, intensified projection direction k5, multiple shelled thermal spheres 7;

Figure 15 spheroid cylindrical energy barrier laced with shelled thermal spheres 10, physical reflection 11, ultimate encompassing focal directional shift l4, final intensified encompassing directional imploding projection shift l5.

Figure 16 illustrates, conception 1 with conceptual position 2 as coincident projections lettered A-C, positional energy barrier 3 as coincident spatial-energy layers lettered a-e, conceptual position through spatial-energy layers Aa2-Ae2, conception projection through spatial-energy layers Aa5-Ae5, projection energy refraction 12 through spatial-energy layers Ba12-Bc12, DNA 13, DNA energy refraction Bd13, Be13, synapse 14, synapse firing Cb14, Ce14 with clear-medium string 15 created at Cd15, Ce15 through 1, alternate focal direction 4 and projection direction 5 channeled through the clear-medium string 15 and energy conduit directed flow towards more physically mirrored environments 16.

As used in the drawing specification, the following reference numerals shall refer to these items:

1 - conception

2 - conceptual position

3 - positional energy barrier

4 - conception focal direction

5 - conception projection direction

6 – shelled ringed thermal reaction

7 - shelled thermal spheres

10 - cylindrical energy barrier laced with shelled thermal spheres

11 - physical reflection

12 – clear-medium projection energy refraction

13 - DNA

14 – synapse (gap between conductive metal disks)

15 – clear-medium string

16 - energy conduit directed flow

a-l - lettered event chain

A-C - lettered clear-medium coincident projections

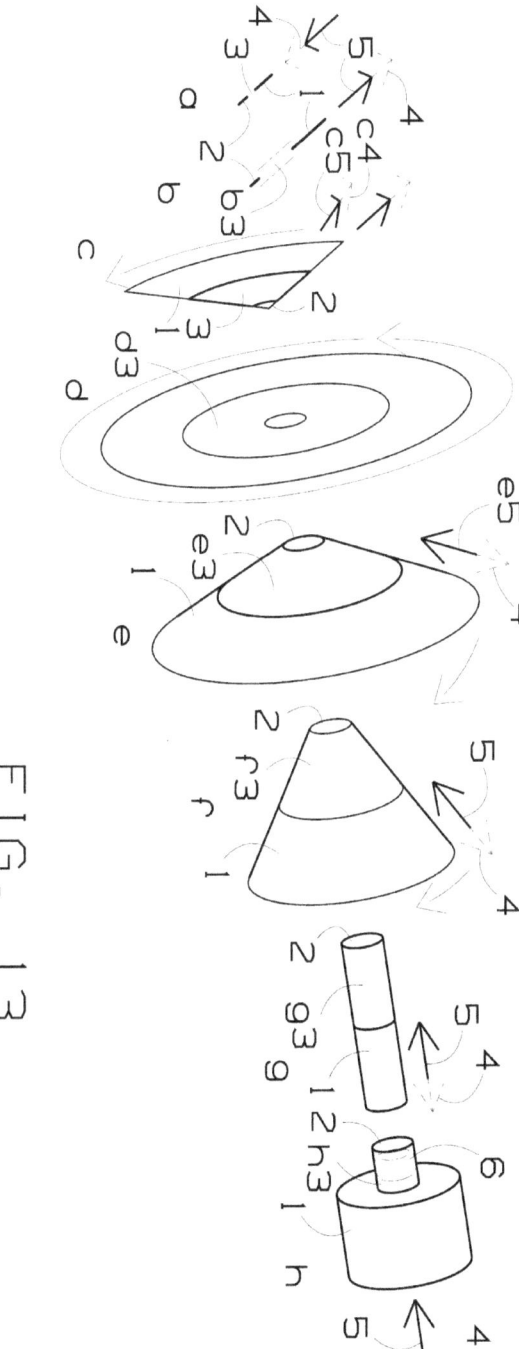

FIG. 13

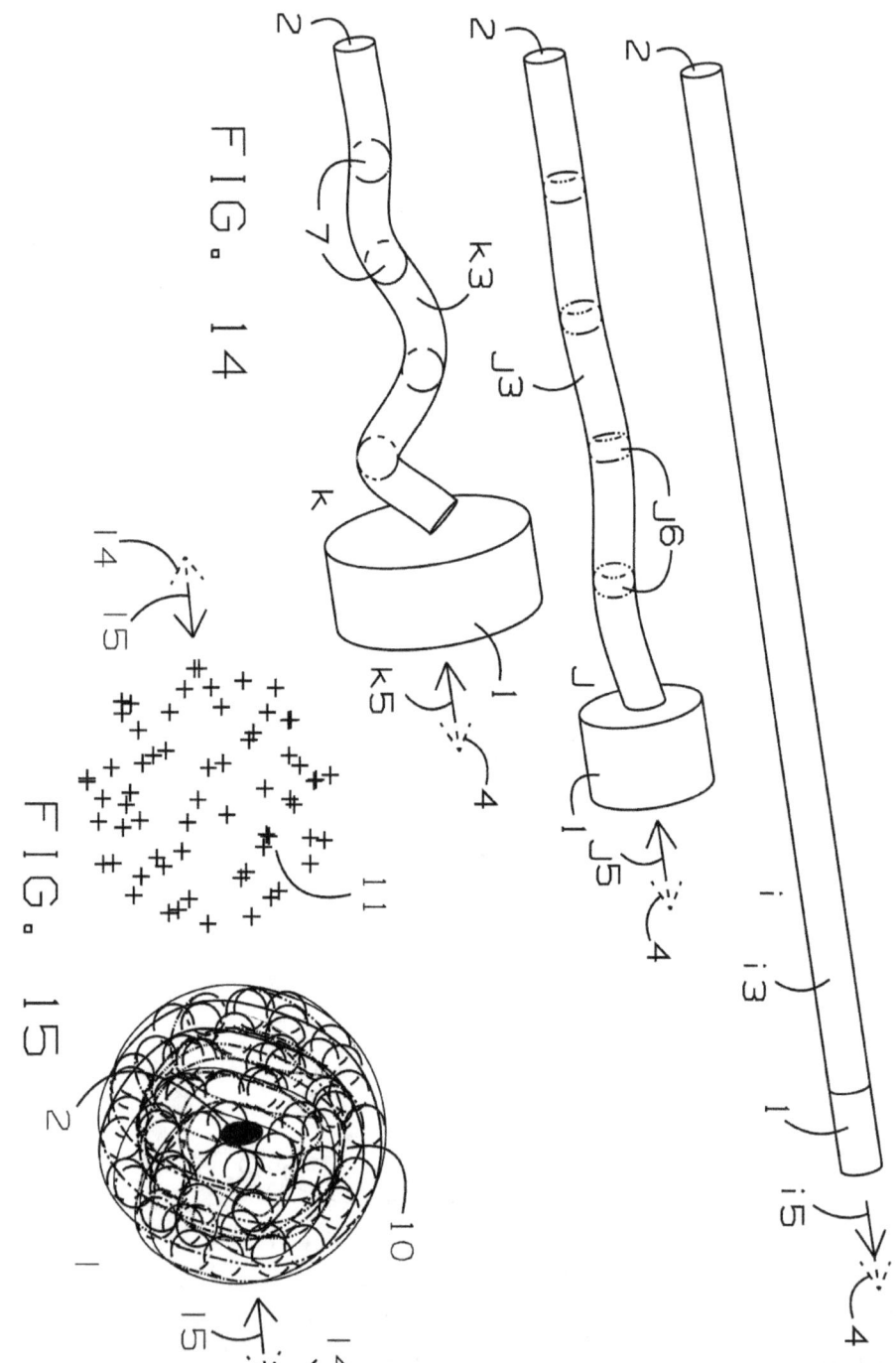

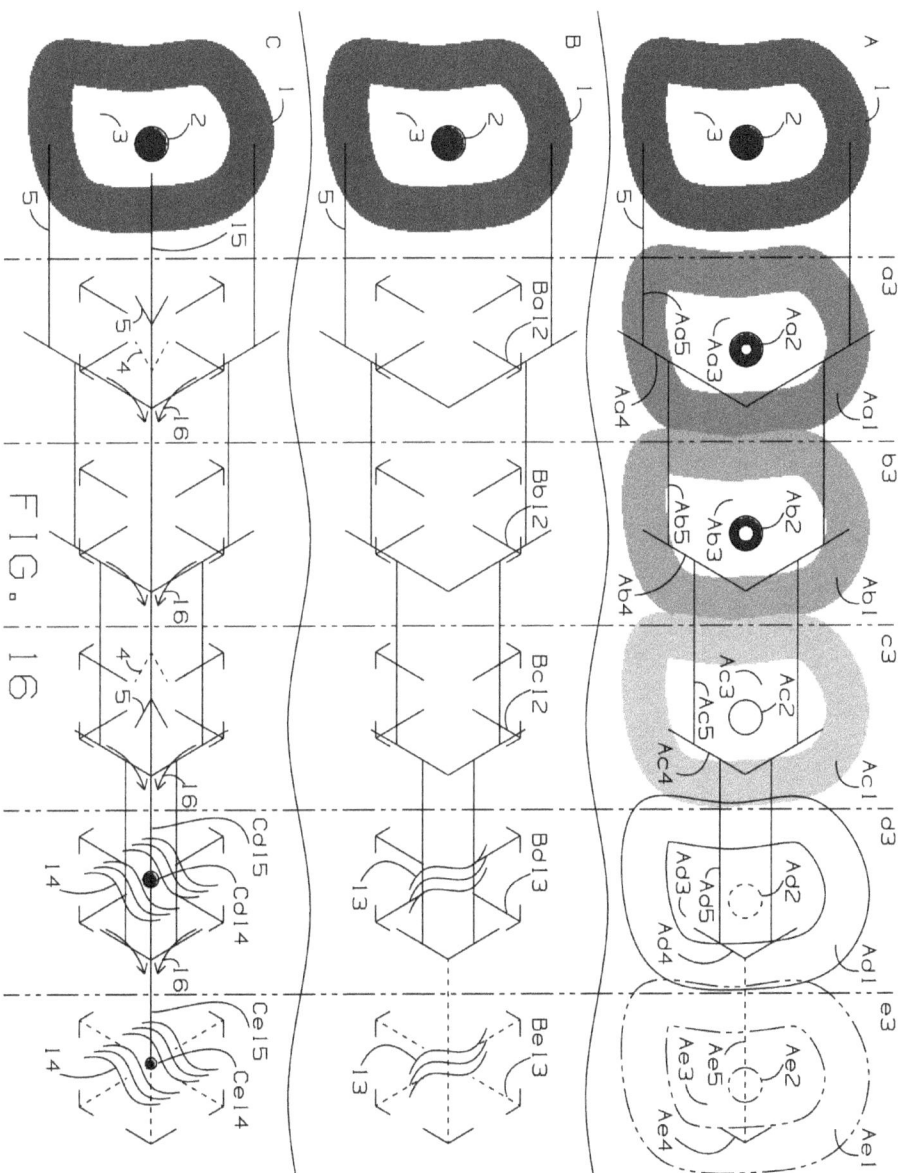

www.ingramcontent.com/pod-product-compliance
Lightning Source LLC
Chambersburg PA
CBHW070428180526
45158CB00017B/923